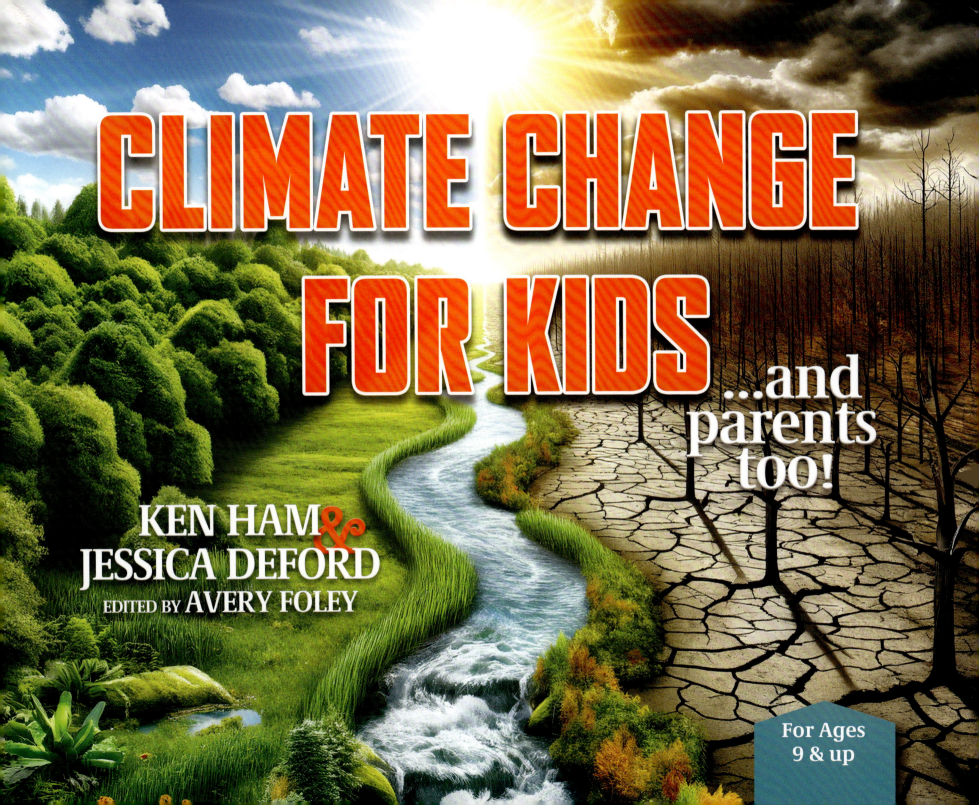

First printing: May 2024
Second printing: June 2024

Copyright © 2024 by Ken Ham, Jessica DeFord and Master Books®. All right reserved. No part of this book may be reproduced, copied, broadcast, stored, or shared in any form whatsoever without written permission from the publisher, except in the case of brief quotations in articles and reviews. For information write:

Master Books, P.O. Box 726, Green Forest, AR 72638
Master Books® is a division of the New Leaf Publishing Group, LLC.

ISBN: 978-1-68344-358-2
ISBN: 978-1-61458-890-0 (digital)
Library of Congress Number: 2024932743

Design by Jennifer Bauer & Diana Bogardus
Cover Design by Diana Bogardus

Scripture quotations are from the ESV® Bible (The Holy Bible, English Standard Version®), © 2001 by Crossway, a publishing ministry of Good News Publishers. Used by permission. All rights reserved. The ESV text may not be quoted in any publication made available to the public by a Creative Commons license. The ESV may not be translated in whole or in part into any other language.

Please consider requesting that a copy of this volume be purchased by your local library system.

Printed in China

Please visit our website for other great titles:
www.masterbooks.com

For information regarding promotional opportunities, please contact the publicity department at pr@nlpg.com.

Table of Contents

Introduction 4
A Faulty Foundation 8
Observational Science 10
The Science of Climate Change 12
Do Climates Change? 16
What Does the Data Say? 18
Are We the Problem? 20
Carbon Dioxide CO_2 22
Just a Natural Cycle? 24
What About Extreme Weather Events? 26
Proxy Data 32
A Warmer Earth? 34
One or Many Ice Ages? 36
Can You Spot the Missing Data? 38
What About the Polar Bears? 40
Will You Be Washed Away? 42
Bleaching by the Beaches 44
We Must Ask "Why?" 46
Is Sustainable Energy Sustainable? 48
Cheap Energy Is Good for Everyone! 50
God's Word Is True! 52
More Importantly...The Biblical Worldview 54
Building the Right Foundation 56
Biblical Climate Ages 58
    Perfect Climate Age 60
    Groaning Climate Age 64
    Flooding Climate Age 66
    Icy Climate Age 70
    Shifting Climate Age 72
    Fiery Climate Age 74
    Heavenly Climate Age 76
Stand Firm 78

*The heart of man plans his way,
but the Lord establishes his steps.*
–Proverbs 16:9

Introduction

From 1971 to 1974 I studied for a science degree at a university in Australia so I could become a science teacher. For my degree in Biology and Environmental Science, I had several courses dealing with how man should interact with the environment. At that time in Australia, I didn't have the opportunity to go to a Christian university and study from a biblical worldview perspective.

My professors taught me from a secular worldview, meaning they started with man's beliefs about the universe and built their thinking to understand the world we live in on that foundation. Because of their foundation of man's word, they started with the view that the universe, including the earth and all life, arose by natural processes — no God was involved — as life supposedly evolved over millions of years by chance and random processes.

The Unisphere Globe in New York

In the secular worldview, man is not special, as we're just another animal. Man doesn't have the right to have dominion over the earth and all living things. There is no such thing as sin, and man is basically good. Because of this evolutionary teaching, I was told that everything natural is good. My professors didn't believe sin had affected the earth in any way. Because they didn't believe that God created man and that man had sinned, they didn't believe in the saving gospel. Their belief was that man could save himself and man could save this planet for the future.

They certainly didn't believe what God said in Genesis 1:28, *"Be fruitful and multiply and fill the earth and subdue it, and have dominion over the fish of the sea and over the birds of the heavens and over every living thing that moves on the earth."*

By Ken Ham

The textbooks I had been given stated that humans were producing too many people and this would cause overpopulation and destroy the earth within 10 years ... but that never happened.

I was told that the ice caps were melting because of all the pollution man was putting into the atmosphere and this would flood the continents and cause great devastation ... but this never happened.

I was told we were running out of oil and gas and so we had to stop using gasoline cars ... but then we found there was plenty of oil and gas to last probably hundreds more years.

I was also told that because the earth and life had evolved slowly over millions of years, if man made any negative changes to the planet, then it could take nature a very long time to repair it.

Many years later I saw a movie featuring American politician Al Gore making all sorts of similar predictions about how man was causing climate change and this would devastate the earth ... but his predictions didn't come true either.

We have many politicians and others today claiming man is causing climate change and this will destroy the earth in just a few years. They claim we have to stop using oil and gas as it's polluting the world and is on the verge of causing massive destruction.

CLIMATE CHANGE PREDICTIONS[1]		TRUE ✔	FALSE ✘
1950s	It was predicted Earth would run out of fossil fuels, a concern also echoed in the 1970s		✘
1960s	Antarctic ice sheets would melt from global warming and create a drastic rise in sea levels		✘
1970s	Secular scientists feared a new ice age was beginning, warned of food shortages, erratic weather because of global cooling		✘
1980s	Predictions of massive man-made emissions from cars, factories, or other sources would be catastrophic		✘
1990s	Solar activity cycles linked primarily to the cause of climate change		✘
2000s	Damage to the earth or biomes will take very long periods to be renewed		✘
2013	Predicted global warming per surface temperature data results from a 15-year study		✘

In fact, some such people are claiming the earth will be destroyed in a few years if this supposed man-made climate change isn't stopped. But I have quite a different prediction. I predict that, in another 10 years, we will find these predictions haven't come true either for two reasons:

1. God's Word makes it quite clear that man can never destroy the earth (Genesis 8:22).

2. Real observational science (not the radical interpretations of some of the data) does not support the alarming claims of these activists.

▶ **As you'll discover in the next section of this book, yes, the climate is changing — but it's not your fault and it won't destroy the earth!**

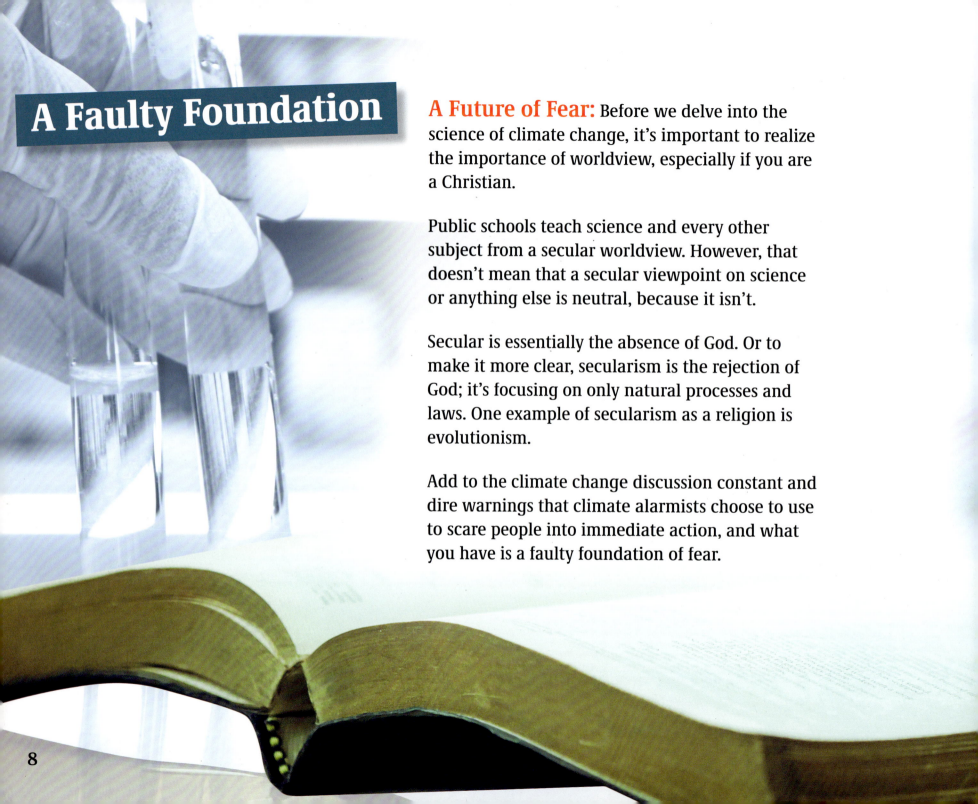

A Faulty Foundation

A Future of Fear: Before we delve into the science of climate change, it's important to realize the importance of worldview, especially if you are a Christian.

Public schools teach science and every other subject from a secular worldview. However, that doesn't mean that a secular viewpoint on science or anything else is neutral, because it isn't.

Secular is essentially the absence of God. Or to make it more clear, secularism is the rejection of God; it's focusing on only natural processes and laws. One example of secularism as a religion is evolutionism.

Add to the climate change discussion constant and dire warnings that climate alarmists choose to use to scare people into immediate action, and what you have is a faulty foundation of fear.

▶ **This fear has also created another phenomenon – "eco-anxiety."**

Christians have no need to fear. We know God and what He has told us in the Bible. Discover the truth of science, climate change, and the true history of the world in this powerful exploration of climate change science and the importance of developing a strong biblical worldview.

Observational Science

By Jessica DeFord

I attended a secular university to obtain my undergraduate degree in a science field starting in 2010 — about 40 years after Ken Ham was pursuing his science degree at a university in Australia.

Even though the start of my university career was many years later, I was being taught the same evolutionary ideas and claims that the earth would end in destruction due to climate change. (I even saw the Al Gore movie Ken mentioned while I was in high school!)

Throughout my education, I was also told man is destroying our planet. That if we don't act quickly, earth as we know it will end. Images of melting ice sheets, dying polar bears, damaged coral reefs, and extreme weather events in my textbooks painted a bleak picture. This instilled a lot of fear and anxious thoughts for the future. This even continued after my education when I started my career as a research scientist. At the federal and state agencies I worked for, climate change was touted as the driving force of ecological destruction.

Climate change was a nail that was continuously hammered into the foundation of my education and science career. I had heard it so often that I believed what I was taught about it. And as a young person I didn't second-guess my professors or coworkers when they agreed with it. But the irony was that I didn't even know what climate change was.

Sure, I had seen a documentary about it, listened to my professors as they lectured about it, and even nodded in agreement with my coworkers who discussed it. But if you would have asked me to explain it, I wouldn't have been able to. My only defense would have been to tell you that it was the reason for so much destruction on the earth and that it was a problem.

▶ **When climate change panic is induced and alarms are sounded in the media or in the halls of academia, we must exercise discernment.**

God's Word, in Proverbs 17:24, says, *"The discerning sets his face toward wisdom, but the eyes of a fool are on the ends of the earth."*

I probably would have rambled on about the harmful carbon emissions man was spewing into the atmosphere or said something about how the sea ice was declining, decimating polar bears. I could only repeat the rhetoric that I had been told about climate change, because I didn't know any of the science behind it.

As a science student, one of the first things I was taught was to ask questions and to critically think about the information that was presented to me. But many of my professors presented data during lectures as if the science was an undisputed fact. This is also how scientific data about climate change is often presented.

To understand climate change and other life events, you need to use a biblical worldview. In this part we will take a look at the science behind climate change. When we study the observational science, we can see that science does not support the claims regarding man-made climate change and we can stand firm on the authority of God's Word when we are bombarded with claims of the earth's destruction.

From God's Word

PROVERBS 17:24

24 The discerning sets his face toward wisdom, but the eyes of a fool are on the ends of the earth.

BE WISE!!

The Science of Climate Change

So, what is climate change? Well, we can define it in the first part of this book as "significant and long-lasting change in the Earth's climate and weather patterns."

What is climate and weather?

1. CLIMATE is the average weather conditions over a long period of time.

2. WEATHER is the current atmospheric conditions such as the temperature, wind, or precipitation (rain or snowfall). This changes on a day-to-day or even an hourly basis.

Now put climate and weather together: the long-term trends in temperature, precipitation, or other weather patterns in a specific area contribute to that area's climate.

So a dramatic or long-lasting change in weather and therefore climate is what it means to have "climate change."

Much of what you read about climate change is about how humans are causing it, or at least making it a lot worse. The examples on the following page are often blamed for causing climate change.

However, there are many other factors, including natural cycles of the earth and sun, that impact the climate.

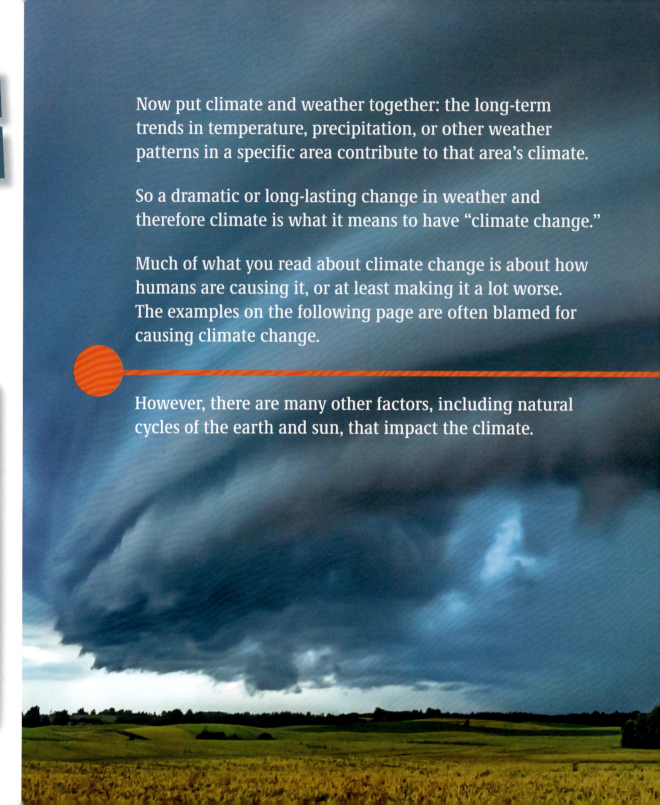

IT'S COMPLICATED

Earth's weather and climate isn't as simple as just temperature and precipitation. Other factors play a part in why certain areas might have warmer or colder weather and climate than others.

For example, if you travel to the Arctic Circle, the climate will be much colder than if you were to travel to the tropical islands of Hawaii. Because a region's latitude (that's the distance north or south from the imaginary circle around the middle of the earth, called the equator) will affect its climate.

There are many ever-changing processes that contribute to a region's climate. Thankfully, many of these processes are cyclical, reoccurring in a predictable pattern of months, years, or even decades. It's all part of God creating an orderly earth and universe that we can study.

And that's not all. Wind, ocean currents, cloud patterns, and more all affect climate.

Cloud patterns → **Wind** → **Ocean currents** → **Solar activity**

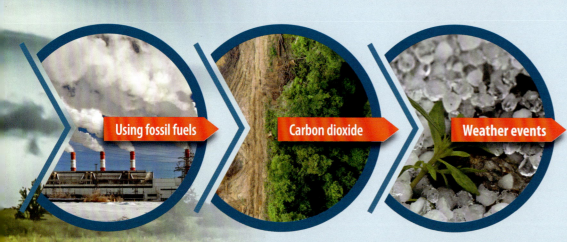

Using fossil fuels → **Carbon dioxide** → **Weather events**

The use of fossil fuels, increases in CO_2 concentration, and increased weather events are all reasons climate alarmists claim humans are causing man-made climate change. But there are many dynamic, cyclic processes that God designed that contribute to changes in climate. God is both the author and sustainer of His creation. *"The earth is the LORD's and the fullness thereof, the world and those who dwell therein,"* Psalms 24:1 affirms.

NORTH POLE

Close to the equator?
Crank up the air conditioning! A region's latitude (the distance north or south from the equator) will affect its climate.

Tilted Axis
The seasonal tilt of the earth can also affect an area's climate because it means the sun's rays will reach the earth at different angles and there will be variation in the length of a day depending upon the time of year or the earth's rotation.

EQUATOR

POLAR AXIS

VERTICAL AXIS

Closer to the poles?
Grab a coat and flashlight! The North Pole is light for about six months in summer and dark for six months in winter.[2] The South Pole is the exact opposite of this.

SOUTH POLE

Do Climates Change?

Secular scientists and creationists agree that climates do change. The question isn't "do climates change?" (Yes, they do). The question is "are we causing it, and does it matter?"

Ever heard of "global warming"? While it's not the preferred term today (climate change is), it's what most people actually mean when they say "climate change."

This warming is what so many people fear. They believe that human activity, such as industry, manufacturing, and burning fossil fuels (like oil and gasoline) since the Industrial Revolution of the 1800s, is causing Earth's temperature to rise. They believe if we don't bring down this "fever" soon, Earth, and all the life that lives here, is doomed.

▶ **Global warming is the overall warming of the earth's surface over an extended period of time.**[3]

To know if the earth is warming or not, scientists measure the earth's global surface temperature. That's not an easy task either (the earth is rather large, after all!). To do it, scientists have been taking land-based temperature measurements at weather stations since 1880. But there are some problems: these weather stations are not evenly spread out around the earth, and day-to-day weather changes at the station locations can limit the data that is collected.

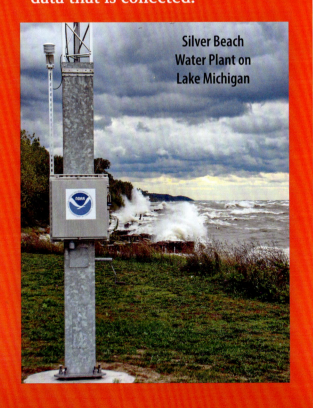

Silver Beach Water Plant on Lake Michigan

So, in an attempt to get a better data set for monitoring global temperature, scientists also began to monitor global temperature by satellites in 1979. **That means we only have about 140 years of temperature data,** and only about 40 years with more accurate global data!

NASA Satellite Earth-Focused Climate Observations[4]

Years	Satellite
1964 to 1970s	NIMBUS Series satellites
1970s to currently	LANDSAT Series satellites
1999 to 2013	ACRIMSAT
1999 to currently	QuikSCAT
2001 to 2013	Jason-1
2001 to 2006	SAGE III Meteop-3M
2003 to 2009	ICESat Satellites
2003 to 2020	SORCE
2007 to currently	TERRA Satellites
2007 to 2021	AQUA Satellites
2007 to 2023	AURA Satellites
2008 to 2019	Jason-2 (OSTM)
2016 to currently	Jason-3
2018 to currently	ICESat-2
2019 to currently	NASA-ISRO

What Does the Data Say?

The observational data shows that the global surface temperature of the earth has been warming over the past 100 years or so since it has been recorded. (Remember how we said that, yes, climates do change.) But should we panic over this warming? Consider this: we know climates are cyclic and ever-changing. So, let's do what scientists do: ask some discerning follow-up questions.

Let's Ask

1. How much has the global surface temperature increased?
2. Is the increase in warming part of a natural climatic cycle?
3. Is global warming occurring because of human activity?

How Much Warming?

The amount of global surface temperature warming that has happened over the past 100 years is estimated to be between 2.7 to 3.6 degrees Fahrenheit (1.5 to 1.8 degrees Celsius). But this warming estimate didn't come solely from the observational data collected at weather stations and by satellites. It's based on computer models.[5] What you input into these models will decide what predications the computer model provides.

So the predictions don't necessarily reflect the real-world, observational data. And one study that compared computer climate models to the observational data found every single climate model they studied overpredicted warming! The models didn't match the actual data. Computer models can be good, but they aren't perfect, and the biases or the quality of data the computer receives will impact its predictions.[6]

▶ **So how much has the earth warmed since the 1880s? We don't really know!**

But wait ... there's more! Climate researchers generally assume Earth maintains a constant average temperature and that our atmosphere traps more heat from the sun than what is returned to outer space. But if either of those assumptions are wrong, (and there's evidence that both are wrong!) the models will make wrong predictions.

Climate change alarmists often use this kind of image to make it seem like excessive warming is occurring very quickly.

Are We the Problem?

Now we need to talk about greenhouse gases. Human activity, primarily the burning of fossil fuels such as petroleum, coal, and natural gas, is seen as the cause of climate change (global warming). In other words, we're the problem.

Just like the glass panels of a greenhouse prevent heat from escaping, greenhouse gases, such as carbon dioxide (CO_2) are believed to trap infrared energy (heat) from escaping into outer space. Since humans burn fossil fuels and burning fossil fuels produces CO_2 and CO_2 traps heat ... we must be responsible for any warming ... right? Not so fast!

▶ Climate activists say carbon dioxide is the primary greenhouse gas.

Carbon dioxide is a greenhouse gas, but it's not the only one.[7] Our atmosphere was designed by God with many other greenhouse gases such as methane and water vapor.

Water vapor is the most abundant greenhouse gas and accounts for somewhere between 60–95%[8] of the greenhouse gas effect. So, CO_2 is not the most important greenhouse gas! Water vapor is. So, no, we are not causing the warming we've observed in the past 100 years.

Did you know that carbon dioxide levels have only been continuously monitored since the 1950s? That's only 70 years of data![9] Past carbon dioxide levels are inferred from tree ring and ice core data based on wrong evolutionary dates of hundreds of thousands of years.

Examples of water vapor

Carbon Dioxide CO_2

Carbon dioxide is a greenhouse gas, but is it bad? Many climate activists want you to believe that it's a harmful pollutant, but did you know that earth can actually benefit from an increase in CO_2?

Carbon dioxide in the air is critical for keeping life alive on earth. God has designed a process called **photosynthesis** where plants take in carbon dioxide, water, and sunlight and convert them into sugars and oxygen. We (and other living things) breathe in oxygen and release carbon dioxide. It's an amazing cycle designed by an all-wise God.

Since plants need CO_2 → more CO_2 means
- → More plants
- → Healthier, less stressed plants
- → Plants that are able to hold water more efficiently, increasing the amount of food they produce

→ Energy from light turns to chemical energy in the form of glucose (sugar) and oxygen being released.

Water + Carbon Dioxide	leads to →	Glucose + Oxygen

The Process of Photosynthesis

WATER ABSORBED THROUGH LEAVES AND SOIL CO_2 GOES IN AS WELL

GLUCOSE (SUGARS) & O_2 ARE PRODUCED BY PLANTS

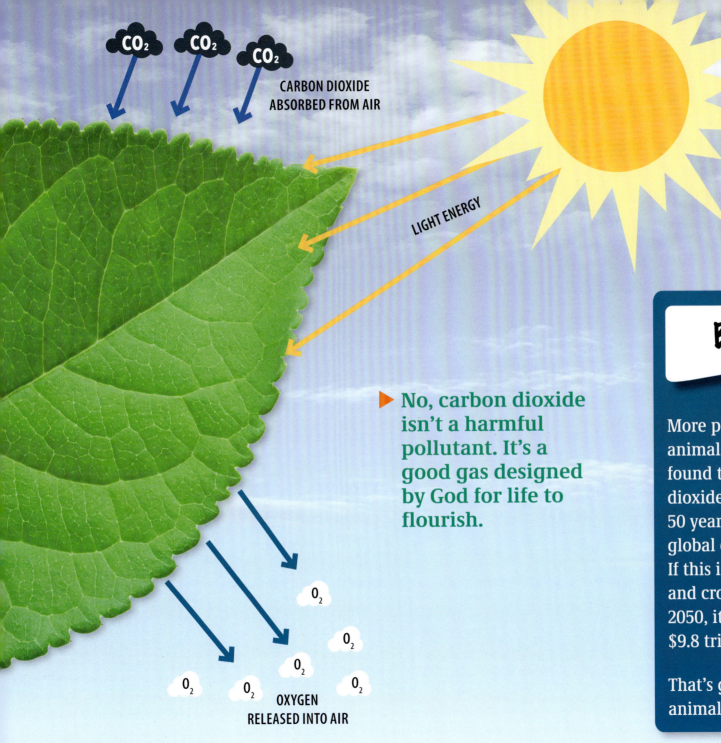

CO₂ CARBON DIOXIDE ABSORBED FROM AIR

LIGHT ENERGY

▶ No, carbon dioxide isn't a harmful pollutant. It's a good gas designed by God for life to flourish.

OXYGEN RELEASED INTO AIR

Bottom Line

More plants equal more food for animals and humans. One study found the increase in carbon dioxide in the atmosphere over 50 years (1961 to 2012) increased global crop yield by $3.2 trillion.[10] If this increase in carbon dioxide and crop yield continues into 2050, it would add an additional $9.8 trillion worth of food.[11]

That's great news for both animals and people!

So, if humans aren't causing the warming by burning fossil fuels, who or what is responsible? A great place to start the search is with natural cycles that we know influence weather and climate.

Just a Natural Cycle?

Let's consider La Niña and El Niño, two climate patterns that globally impact ecosystems and weather. La Niña causes cooling while El Niño causes warming. Climate alarmists are often concerned with added warming, but El Niño is part of a normal cycle that naturally makes things warmer. In fact, the warm water can affect the weather for several months or even years, further raising the water temperature and dumping heavy rainfall. El Niño takes place every 3–7 years.

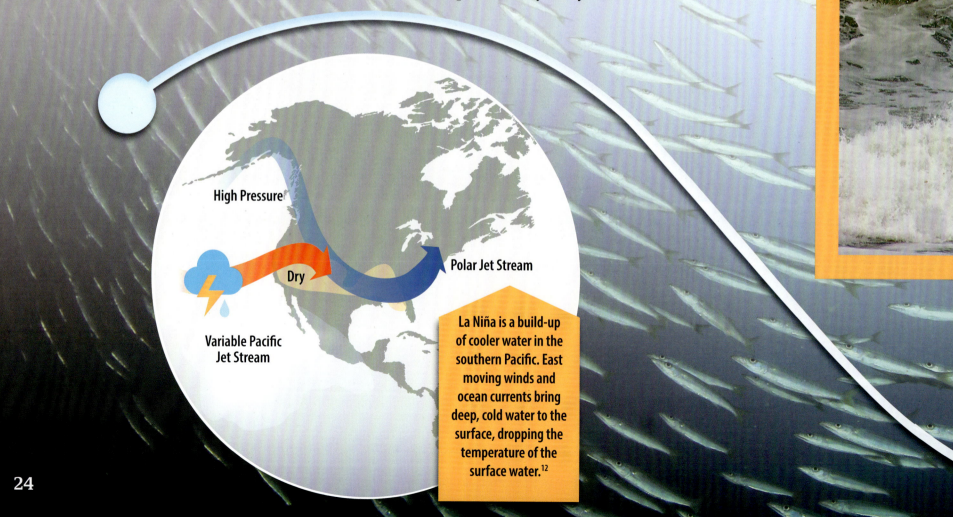

High Pressure

Dry

Polar Jet Stream

Variable Pacific Jet Stream

La Niña is a build-up of cooler water in the southern Pacific. East moving winds and ocean currents bring deep, cold water to the surface, dropping the temperature of the surface water.[12]

When these events are included in climate models, there's much more agreement between the model's predictions and the observational data. This means we can't leave out natural cycles like these when discussing climate change – they make a big difference!

El Niño is a body of warm water located in the western area of the Pacific Ocean.[13] It is believed to be caused by trade winds that blow from east to west across the Pacific.

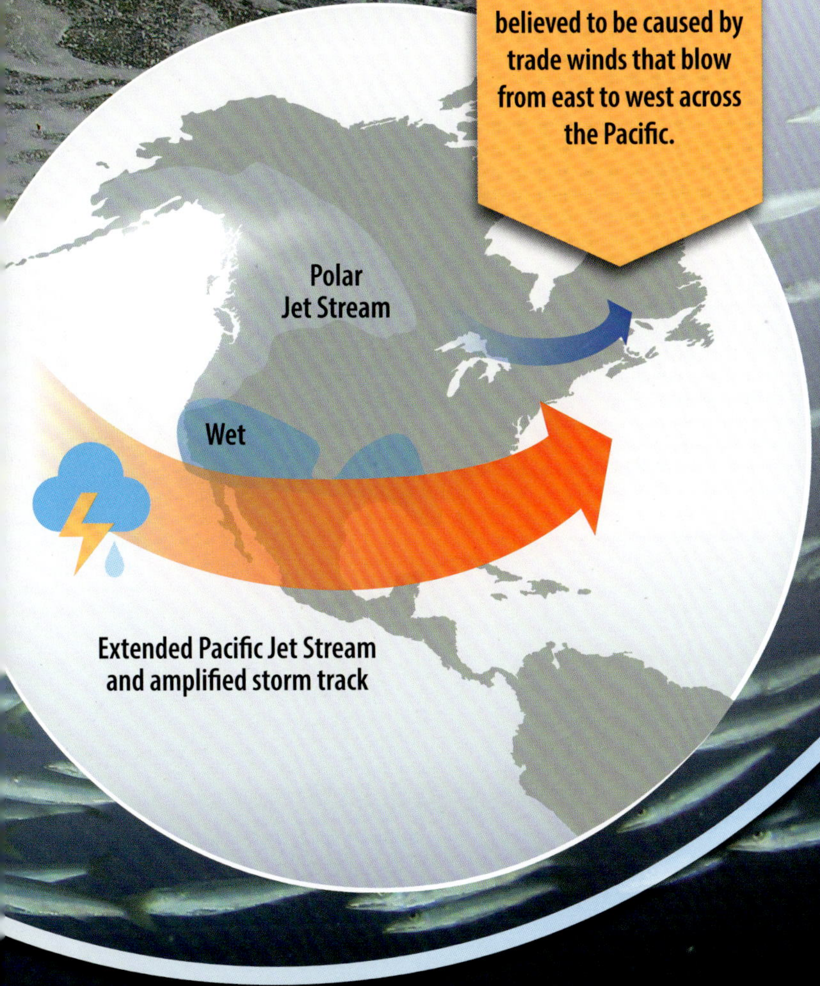

Polar Jet Stream
Wet
Extended Pacific Jet Stream and amplified storm track

USA
THE BAHAMAS
CUBA
Turks and Caicos Islands (UK)
Aruba (NL)
COLOMBIA

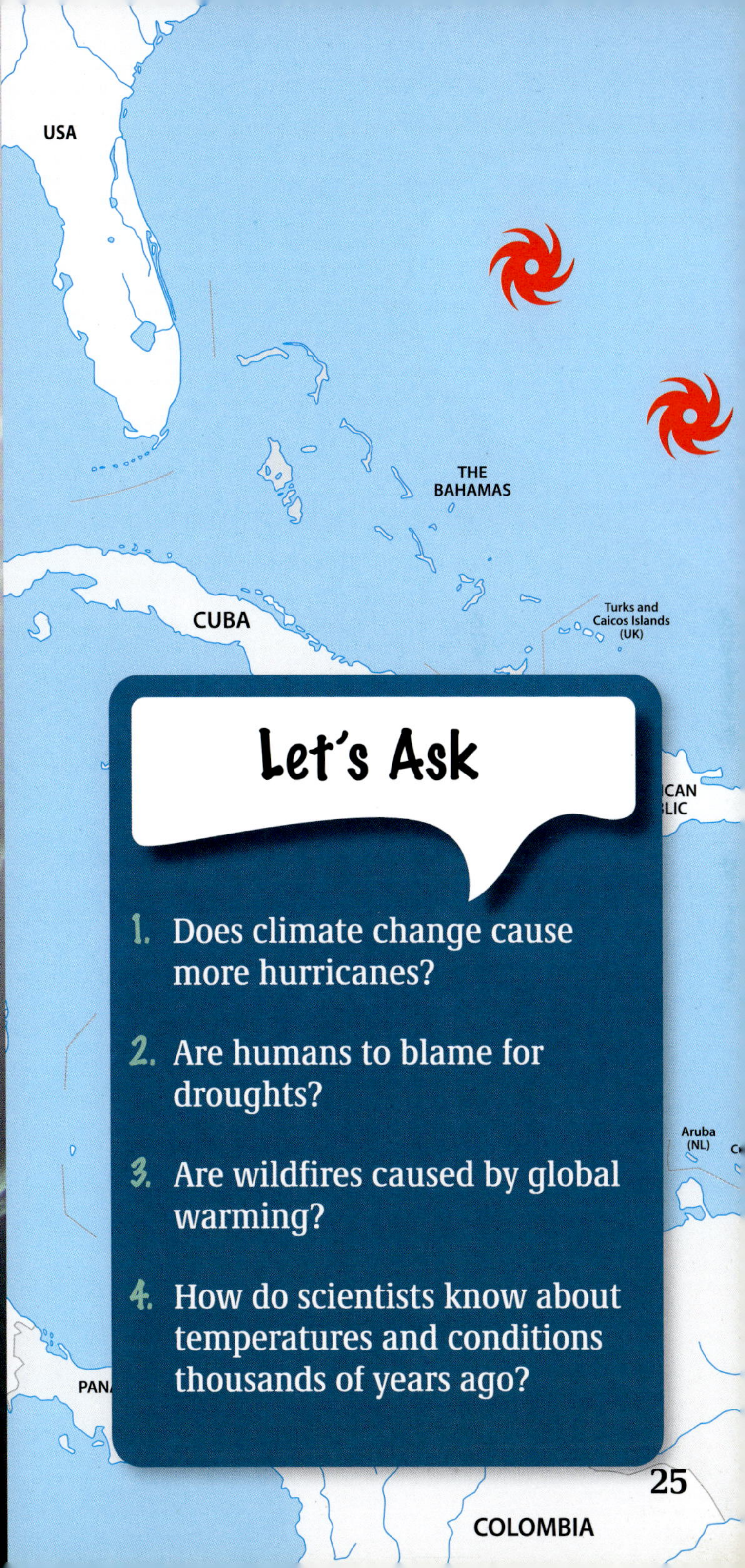

Let's Ask

1. Does climate change cause more hurricanes?
2. Are humans to blame for droughts?
3. Are wildfires caused by global warming?
4. How do scientists know about temperatures and conditions thousands of years ago?

What About Extreme Weather Events?

Every time there's a hurricane, wildfire, or drought, man-made climate change is blamed for it. But are we really causing these extreme events?

▶ **First, we must remember that God is both author and sustainer of all of His creation.**

His Word tells us that He is in control of the weather when He proclaims in Genesis 8:22.

When Adam and Eve sinned in Eden, the world was no longer perfect. That meant that the perfect weather God had created is now broken and different. Now let's examine what the observational data shows us regarding weather events.

From God's Word

GENESIS 8:22

DO NOT FEAR!

²² While the earth remains, seedtime and harvest, cold and heat, summer and winter, day and night, shall not cease."

Hurricanes: Lots of people blame climate change for the hurricanes that have slammed America in recent years. But there's actually little evidence that human activity has increased the number or severity of hurricanes. Actually, a database of the National Oceanic and Atmospheric Administration (NOAA) that has hurricane records going back to 1851 reveals there's been a 50% reduction in the number of major hurricanes in the United States that have made landfall since the 1930s. That means you were much more likely to see a hurricane 100 years ago than today.[14]

Why then?

Why are hurricanes today causing billions of more dollars in damage? Well, more people are living in the United States now than 100 years ago and there are far more buildings, and larger, more expensive homes, built on beaches, the sea shore, and known flood plains. Urbanization (converting farmland into a city) also creates problems for water runoff, causing extra flooding.

Three Category 5 hurricanes that have impacted the United States and caused billions of dollars in damage. →

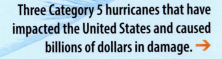
Hurricane Ian
September 28, 2022

Hurricane Katrina
August 29, 2005

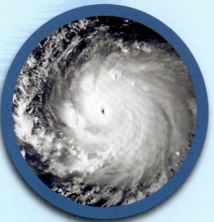
Hurricane Hugo
September 15, 1989

Droughts: What about severe droughts — are they caused by climate change? If so, we should be experiencing the worst droughts of history.

▶ **And yet historical records indicate that "mega-droughts" occurred before the 14th century.**[15]

Current monthly data from the National Oceanic and Atmospheric Administration that shows which parts of the United States are experiencing very wet, flood-like conditions or very dry, drought-like conditions reveals that, if anything, it appears we're currently experiencing slightly less droughts, not more. This is true around the world. The actual historical and current data suggests that the earth experiences natural cycles of wet and dry conditions.

From God's Word

JEREMIAH 17:7-8

7 "Blessed is the man who trusts in the LORD, whose trust is the LORD.
8 He is like a tree planted by water, that sends out its roots by the stream, and does not fear when heat comes, for its leaves remain green, and is not anxious in the year of drought, for it does not cease to bear fruit."

TRUST IN GOD!

We have artifacts that record droughts in human history such as the Famine Stela (below). Droughts are also noted in the Bible — Jeremiah 14, Genesis 12, and Ruth 1 are some examples.

God-designed, natural processes on our world can sometimes lead to droughts in our fallen world.

Mega-blazes are fires that destroy unusually large areas and are very intense. Because of concerns with the climate, people fear there will be more of these occurring.

Wildfires: When wildfires rip through an area, destroying forests and homes, climate change is blamed. But fire isn't actually bad! Periodic fires are important for the health of our forests, so much so that God created many creatures that rely on fire to reproduce or continue their life cycle.[16] When carefully managed so they don't get out of control, prescribed (man-made) fires can help natural ecosystem cycles. But what about out-of-control "mega-blazes"?

▶ **Climate change isn't to blame. Sadly, several poor government policies are largely at fault.**[17]

From God's Word

ISAIAH 43:2

2 When you pass through the waters, I will be with you;
and through the rivers, they shall not overwhelm you;
when you walk through fire you shall not be burned,
and the flame shall not consume you.

NOT CONSUMED!

30

Controlled burns are essential to manage underbrush. These fires are also known as prescribed fires and are focused on the health of the forest and the safety of people living nearby. This effort gets rid of invasive species, enriches the soil, and helps to limit fuel for wildfires.

To try to stop carbon dioxide from entering the atmosphere, policies have been put in place to reduce the number of controlled, prescribed burns. But without occasional fires, too much wood fuel (dried trees and leaf litter) accumulates, and when a lightning strike or human carelessness causes a fire it spreads too quickly and gets out of control.[18] Instead of worrying over carbon dioxide, people must be wise stewards of God's creation, exercising dominion to prevent dangerous wildfires while also managing them so they can benefit the ecosystem.

Why then? Fire is how forests renew themselves, and can be caused by lightning, though most happen from human carelessness. Fire helps to renew the forest by burning debris, creating areas of sunlight where small vegetation can grow, limiting diseases, and improving the condition of the forest habitat for birds, animals, and insects.

Proxy Data

The dark band in the ice core pictured here is a layer of volcanic ash.

Scientists use the width of tree rings to determine what weather was like in an area in the past.

Much of the climate panic isn't really about the current data but actually about what scientists believe about the past. Those who start with evolution believe they can know the conditions in the distant past and they believe those conditions have remained fairly constant for at least the past 10,000 years (but we know, starting with God's Word, earth is only 6,000 years old). So, to them, any trend in warming must be caused by our actions, not normal climate cycles.

But where does this data about the past come from? Most of it is an interpretation of proxy data.

Proxy data is predictions that are made from past events.

Scientists cannot 100% accurately describe past events if they were not there to directly observe them.

▶ **You can't directly test, observe, or repeat the past!**

So, this proxy data will be filled with assumptions about how old the earth is, what the earth used to be like, and more. With so many assumptions (most of which are not based on God's Word), proxy data can be very misleading.

Researchers take ice samples in the Canada Basin, Arctic Ocean

Ice cores and their air bubbles are an example of proxy data. Others include sediments from the ocean or lakes, tree rings data, fossil pollens, and rare historical documents.

A Warmer Earth?

Now while most evolutionists believe our climate has been stable since the end of the supposed last ice age, they believe that many, many years before that earth was likely much warmer than it is today and the atmosphere had more carbon dioxide.[19] This warmer environment was actually our climate just a few thousand years ago, before the global Flood — not millions of years ago.

And we've had warm periods since the Flood but before the Industrial Revolution, blamed for today's supposedly dangerous warming. The Medieval Warm Period (A.D. 900–1300) is an example. During this period of warming, Vikings farmed on Greenland! Those same areas are now covered in ice. Obviously, our climate has varied throughout history for various reasons.

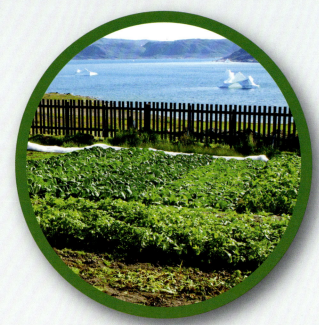

An archaeological site in Hvalsey, Greenland. Scientists seek to find clues to past climate history.

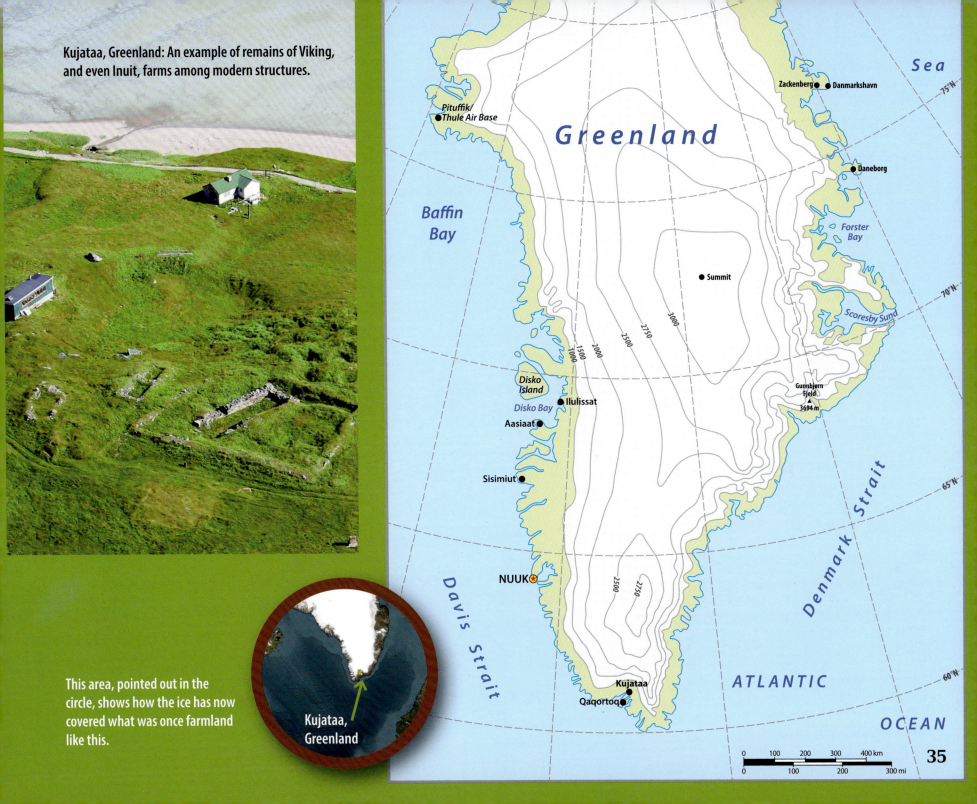

Kujataa, Greenland: An example of remains of Viking, and even Inuit, farms among modern structures.

This area, pointed out in the circle, shows how the ice has now covered what was once farmland like this.

Kujataa, Greenland

One or Many Ice Ages?

The Icy Climate Age was a single ice age that followed the global Flood just a few thousand years ago.[20] This ice age created glaciers that covered around 30% of the Northern Hemisphere (you can still see what's left of the glaciers today in the far north).

Evolutionists also believe in an ice age, but they believe that there have been several ice ages occurring over millions of years. And this belief has a big impact on how they view climate.

Scientists extract ice core samples (long pillars of ice) from glaciers and ice sheets left over from the ice age. They use the data collected from air bubbles and dirt trapped in these cores to make predictions about past climate. While scientists may be able to take some accurate measurements of the ice layers, such as the ice depth or the type of sediment trapped inside, they have to make assumptions about what occurred when the layers were formed because they weren't there in the past to directly observe the ice cores being formed.

Glacier Girl, the WWII plane recovered from almost 300 feet of ice. Secular scientists claim to have discovered 110,000 annual layers in the Greenland ice sheet. By assuming each layer represents a year, relying on their flawed ideas of a very, very old earth, they discount multiple issues that could confirm a young earth. What if each layer is not representing a year, but a weather event adding rain, snow, or ice to the ice sheet during a year or longer? These scientists are blinded by their own assumptions and errant worldviews.

85 m (about 300 ft)

One of those big assumptions is that each layer of ice represents a single year. Because there are many of these "annual layers," scientists believe they give us climate data dating back hundreds of thousands of years. But weather changes constantly and climate doesn't stay the same either, so a layer does not necessarily represent a single year.

Consider this: During World War II (1939–1945) eight planes heading for Scotland through Iceland hit a major storm. Screaming winds, heavy snowfall, and almost empty fuel tanks made it impossible to keep flying, so the pilots were forced into an emergency landing on the ice in a remote area of Greenland. The pilots were rescued but had to leave their planes, abandoned to the weather. Decades later (in the 1980s), a recovery team set out to find and restore these pieces of history — but where were they? The planes had disappeared, buried under more than 250 feet of ice and snow. These layers had formed in less than 50 years!

▶ **Since the assumptions about Earth's age and ice cores are wrong, the interpretations of the data, leading to climate panic, are also wrong.**

So, the assumption that ice layers form only one per year is wrong and these planes proved it! Ice layers form when snow and ice thaws and then refreezes. Many layers can form in just one year. Isn't that really cool?

Can You Spot the Missing Data?

You know what else impacts climate and how much ice and snow exists? The sun! During what some call the Little Ice Age, (a cool period from 1300–1800, that followed the Medieval Warm Period), scientists observed a smaller number of sunspots on the sun.

"Sunspots run in cycles. There is the familiar 11-year cycle. Then there is a 22-year cycle, and there is a long period, chaotic cycle that lasts several hundred years. During the Little Ice Age that lasted from about 1300 to 1880, sunspots were at a general minimum, suggesting that effects on the sun caused the Little Ice Age."[21]

Less sunspots allowed many of the glaciers to grow.[22] The Thames River in London froze over and much of Britain experienced food shortages.[23] Any accurate climate model has to take into account the sunspot cycles on the sun.

Sunspots are dark, cool spots on the sun.[24]

▶ Earth also goes through these long-term variations and these natural cycles have nothing to do with being caused by humans.

From God's Word

JOB 37:5-6, 9-10

5 God thunders wondrously with his voice; he does great things that we cannot comprehend.
6 For to the snow he says, 'Fall on the earth,' likewise to the downpour, his mighty downpour.

9 From its chamber comes the whirlwind, and cold from the scattering winds.
10 By the breath of God ice is given, and the broad waters are frozen fast.

[Handwritten notes: GOD'S WORLD = GOD'S RULES]

Climate alarmists blame most climate change on man-made issues, but as we have seen, there are many God-designed, natural forces that, in this fallen world, can adversely impact the climate on Earth, which people have no control over. Such as the Little Ice Age.

The Little Ice Age ... Was Really Little!

When you hear the "Little Ice Age," you might be thinking this was a worldwide event, but it wasn't. This was mainly a period of cooling that impacted the region of the North Atlantic.

There are even differences about when this little "ice age" occurred. Some date it between the 16th to 19th century, while others prefer the 1300s to around 1850.

Whatever the timeframe, it was a regional event rather than a global one.

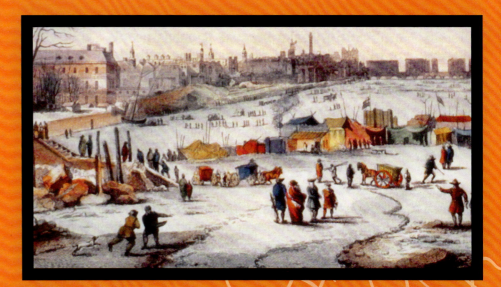

39

What About the Polar Bears?

The glaciers and ice sheets deposited during the Ice Age have been melting and slowly shrinking or sometimes slowly growing since the peak of the ice age. While much of the ice that grew during the Little Ice Age has since melted, research suggests that Antarctic sea ice is expanding,[25] and Arctic sea ice is actually growing a little bit too, compared to the previous 10-year average.[26]

Polar bears travel over massive ranges. When they hunt, they often eat seals because of seal's fat. One seal could provide almost a week's worth of energy for the polar bear.

Supposed man-made climate change is not causing a reduction in polar bear populations.

Now you may have heard that climate change is killing polar bears. They are regularly featured in climate change ads and materials. But polar bears have been doing just fine. Polar bears appear to be thriving.

▶ **In fact, polar bear populations have been slowly growing from around 5,000 bears in the 1950s to likely above 30,000 bears in 2022.**[27]

Will You Be Washed Away?

Flood markers are like rulers that tell you the depth of water during area flooding.

A water level monitoring station with an acoustic sensor.

Many people choose to live and construct homes or resorts near coastal areas.

What about sea levels? Or mega-floods? Will they rise and swallow whole islands, beaches, and cities because of man-made climate change? Like everything we're seen so far, sea level rise actually varies naturally, and scientists monitor water levels with the use of tidal gauges and other technology.

The historical evidence supporting the catastrophic worldwide flood, as described in God's Word, helps us know that we can trust His Word to never flood the entire Earth again.

Data from the National Oceanic and Atmospheric Administration shows that, yes, the oceans have been slowly rising since 1856 but only about 0.113 inches per year[28] – that's not a lot! Even if humans are contributing to this rise (and that's a big if!), it can't be much because the number is already so small.

Even with this information, you can still see predictions about extreme sea levels and how many coastal cities could be impacted or even destroyed by a future date. However, people choosing to live in coastal areas therefore, naturally, run the risk of damage from weather events and increased water levels resulting from those natural disasters.

This is 0.113 of an inch.

Aftermath damage and debris from flooding and storm surge from Matlacha Florida Hurricane Ian Circa October 2022

Bleaching by the Beaches

A rare moment caught on camera when corals under heat stress turn vibrant colors, usually preceding full coral bleaching and death.

Australia is home to the world's largest coral reef, the Great Barrier Reef, home to millions of fish and other sea creatures. But many climate activists believe this beautiful reef is in danger of disappearing due to "coral bleaching."[29]

It is said that there are around 6,000 species of coral[30] in the world today.

Corals and tiny algae normally work together, each helping the other. The algae converts sunlight to food energy for the coral, while the coral is shelter for the algae. When a coral bleaches due to warmer water temperatures, a lack of nutrients, or too much sunlight, they kick out the algae that normally live inside them. When the algae leaves, so does the coral's color, making it appear like it has been bleached white.

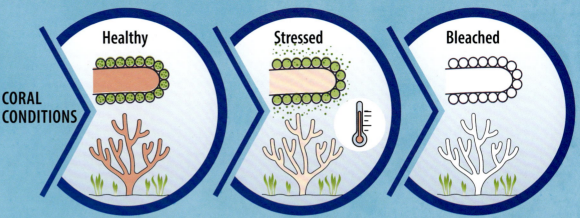

CORAL CONDITIONS

Are humans causing coral bleaching because of climate change? No, this is a natural occurrence (such as after a hurricane) that corals can survive and recover from. While the Great Barrier Reef has experienced some bleaching, Australia's Tropical Marine Research Agency's data set shows that corals are in good health, experiencing record breaking 36-year highs.[31]

This shouldn't surprise us. God has designed coral to be tough, able to deal with a changing world.

God created the sea creatures according to their kinds. Because we still have different sea creatures such as coral today, we know the original kinds must have survived a past major climate event — the catastrophic, global Flood. The coral species we see today are the descendants from the original coral kinds that God created.

Staghorn Coral (Acropora) in the process of bleaching.

From God's Word

GENESIS 1:21

ens." **21 So God created the great sea creatures and every living creature that moves, with which the waters swarm, according to their kinds, and every winged bird according to its kind. And God saw that it was good.** **22** And God

We Must Ask "Why?"

Every time your dad or mom fills the car with gas, you should think of the Flood! You see, the floodwaters dumped mud and minerals in layers, burying millions of plants and animals. Many of those became the fossil fuel deposits we find today which become products we use to drive our cars and heat our homes. When we burn these fossil fuels, they produce a lot of energy, but they also release a lot of carbon. Is this polluting the earth?

▶ Sustainability means sharing or using resources in a way that benefits people and allows those resources to be available for future generations.

Fossil fuels are the results of things that died during the Flood told about in the Bible in Genesis.

Many people believe that it is, saying we need to use only "green" or "sustainable" energy such as wind and solar.

As Christians, we want to use earth's resources sustainably. That's part of being a good steward of creation and loving our neighbor by ensuring future generations have what they need (Genesis 1:28, Luke 10:27). That means that, on the one hand, we avoid extremes that cause destruction to the earth and, on the other, we don't worship the creature/creation rather than the Creator (Romans 1).

God has graciously allowed people to develop technology that has advanced society through the use of different types of energy. A great example is the technological advancement from wood-burning heat to electric, coal, and natural gas (fossil fuels) that helped nations around the world greatly reduce poverty, cold, and hunger.

Now alternative energies such as wind and solar could be used in a similar way if people steward these resources for the glory of God and the benefit of their neighbor. But what happens when these alternative energies are forced on us by governments in a panicked response to human activity supposedly destroying the earth?

Before we jump on the 100% sustainable energy bandwagon, we need to use our discernment and ask,

▶ **"Why is this being pushed so hard and so quickly?"**

Well, many of these types of climate policies are based on climate model data. As we've already seen, any increase in temperature and carbon dioxide because of human activity is really, really tiny, and these climate models don't even properly predict the warming! Wrong or bad data results in wrong models which then results in bad or very rushed policy.

Is Sustainable Energy Sustainable?

Sustainable, or renewable, energy, such as solar, wind, geothermal, and hydroelectric, are considered the only way of the future — and we need to switch right now. But these types of energy are often dependent on environmental conditions such as wind, temperature, and sunlight. That creates a problem — no wind, no energy from wind turbines; no sunlight, no energy from solar panels; really cold or really hot weather, electric cars might not start. New technologies like these can also be very expensive to buy and maintain.

None of these alternative energies are yet as reliable as fossil fuels. Before we force everyone to switch to a specific kind of energy, we need to consider the environmental conditions and how these policies will impact everyday people, especially those in poverty-stricken communities.

Here's an example: In the summer of 2022, legislators asked electric car owners not to charge their cars during heat waves because they knew more people would be using energy to run their air conditioner units.[32] If everyone were forced to own electric cars — which are very expensive — and politicians control energy demand, what would that mean for personal travel or for the supply chain or the economy if businesses also rely on this type of energy for traveling? That could cause huge problems.

Think about this: Now there's lots of wind, sunshine, and heat from the earth — but it takes rare materials such as lithium, cobalt, and copper to make and operate batteries for electric cars, wind turbines, and solar panels. These materials are not abundant, they aren't renewable, and we have to mine them from the earth.

▶ **Is that really renewable and sustainable? Maybe not!**

Cheap Energy Is Good for Everyone!

Did you know that many people around the world still use dried dung (animal poop) and wood for heat? When those are your only energy sources, it's really hard for a nation to lift itself out of poverty. Forcing everyone to switch to much more expensive energies, such as wind and solar, prevents these countries from producing or using their own fossils, or importing them from other countries, to help their own people.

If we're concerned about how environmental problems affect people, like so many climate alarmists claim they are, they should be helping impoverished nations gain access to cheap, reliable energy to allow those people to heat their homes and help them produce infrastructure (buildings, roads, etc.) to withstand natural disasters instead of pushing policies that prevent them from developing.

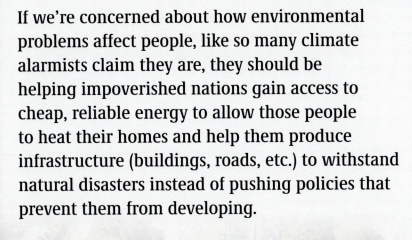

God's Word Is True!

From God's Word

GENESIS 1:31

it was so. ³¹ And God saw everything that he had made, and behold, it was very good. And there was evening and there was morning, the sixth day.

very good!

As we have seen, the observational data confirms the history in God's Word and gives us a framework so we can properly view both past and present climate events. When climate change science is done starting with the authority of God's Word, we also understand how climate policies can affect people made in the image of God.

There will always be new scientific information regarding climates and weather — a new graph, chart, statistic, book, or study to look at. Scientists must remember that science is not the objective standard for truth. Science can certainly inform us, and it does confirm God's Word, but it is not supreme. God and His Word is the ultimate authority by which we must discern all climate and weather information.

After God created everything, he proclaimed it as "very good" (Genesis 1:31). The corruption that we experience in God's creation today is because of man's sin when he disobeyed God in the garden of Eden. Climate alarmists will not solve the problem of man's sin by switching to alternative energy or reducing carbon emissions that they believe are caused by human activity.

It is only through the death of our Lord and Savior Jesus Christ that man will be made new (2 Corinthians 5:17) and exercise wise dominion for the glory of God.

▶ **May we be found faithful to exercise dominion and stewardship over God's creation in a way that brings Him glory and benefits our neighbor.**

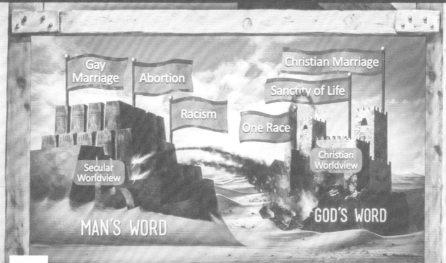

MORE IMPORTANTLY... THE BIBLICAL WORLDVIEW

By Ken Ham

> The modern climate change movement is actually a religion that believes man can save himself and save the planet. Yes, the atmosphere and climate are complex, but the God who made it all and thoroughly understands it all said,
>
> *While the earth remains, seedtime and harvest, cold and heat, summer and winter, day and night, shall not cease* (Genesis 8:22).
>
> We don't need to fear that man will destroy the planet, as God wouldn't let that happen anyway.[33]

> "Instead of arrogantly assuming we know more than God, as humans we must humble ourselves, acknowledge our limitations, and have wise dominion over creation for our good and God's glory. And all that starts with having a biblical worldview of mankind, children, this earth, and our Creator God."[35]

Science can't explain everything. Only the Bible provides the foundation and framework for science and everything else in this world. God's Word is an eyewitness account of history.

> I was interviewed live on CNN by Piers Morgan. The first question he asked me was, "Why don't you believe in climate change?" I answered, "I do believe in climate change." I remember Bill Nye then interjected to claim that I reject climate change. I said I didn't reject climate change and that "there's been climate change ever since the flood."[34]

You have learned a lot about climate change science, so let's do a quick recap.

Building the Right Foundation

Is climate change real? Well, first, we learned and understand what we mean by the term "climate change." According to the Merriam-Webster dictionary, "climate change" is defined as "significant and long-lasting change in the Earth's climate and weather patterns."[36]

So do biblical creationists believe in "climate change"? We sure do, as we have observed significant and long-lasting changes in this earth's climate. We know some areas, like the Sahara desert, used be very lush and well-watered. We know the Vikings used to farm in Greenland, but today the world's largest island is like a frozen wasteland. In fact, the earth has gone through a number of "climate change" events over the past 6,000 years, some more significant than others. And these events are affecting us to this day, and will continue to affect us into the future.

▶ **Without God's Word, one can never fully understand the climate changes we observe.**

Now the only way we would know for sure about what has happened to the climate in the past is if someone was there to tell us the events that occurred. Secular scientists have a problem. They live in the present and have to try to figure out what has happened in the past when they weren't there.

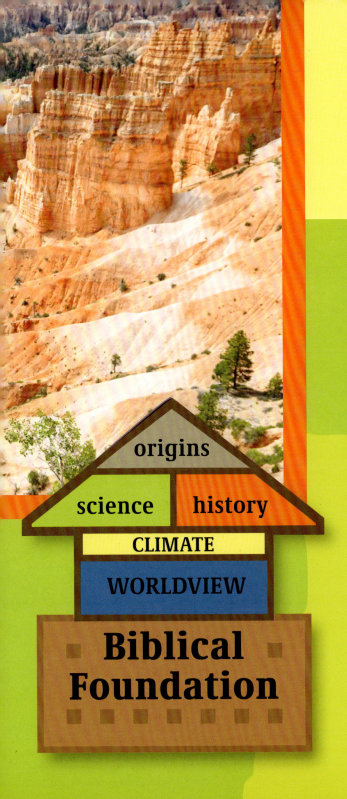

origins

science history

CLIMATE

WORLDVIEW

Biblical Foundation

From God's Word

2 TIMOTHY 3:16

TRUST GOD'S WORD!

¹⁶ All Scripture is breathed out by God and profitable for teaching, for reproof, for correction, and for training in righteousness, ¹⁷ that the

But for the Christian, God, who has always been there, had people record (in the Bible) the history of Earth and the universe so we can know for sure what happened in the past. Once we know the major events of the past, we can begin to properly understand how these events impacted the earth's climate over time, leading up to our present day.

Those who reject God have an enormous problem. Because they reject God's Word concerning the past, they cannot accurately understand why the present world is the way it is and what will happen to it.

I give God's book, the Bible, a special name. I call it "the history book of the universe." So, we are going to use God's history book (the only 100% accurate history book) to help us understand events that have affected climate in the past and how that affects the earth today. Then we will truly understand climate change from a biblical worldview perspective. It's also important to understand that what we believe about climate change will affect what technology we develop and what policies governments will make that can affect billions of people in regard to their quality of life. So these issues are very important!

How should we look at these issues as a Christian to find the truth? Start with a biblical worldview rather than a secular one.

Biblical Climate Ages

So we let God tell us the key information we need to then build our worldview on that foundation as we consider this subject. Actually, we should do that for every subject if we want to truly have a biblical worldview in every area! But let's deal with this one topic: climate change.

As I've read the Bible to understand "climate change" events that affect us to this day, God has revealed **7** climate ages divided into the history in the Bible.

Perfect
CLIMATE AGE – P. 60

Groaning
CLIMATE AGE – P. 64

Flooding
CLIMATE AGE – P. 66

▶ **WORLDVIEW** means your way of thinking about everything. For Christians, our worldview is founded in God's Word.

Icy
CLIMATE AGE – P. 70

Shifting
CLIMATE AGE – P. 72

Fiery
CLIMATE AGE – P. 74

Heavenly
CLIMATE AGE – P. 76

Let's look at each of these **7** climate ages in detail.

Perfect
CLIMATE AGE

- Day One
- Day Two
- Day Three
- Day Four
- Day Five
- Day Six
- Day Seven

When God finished creating the universe, including the earth and all life, He said that everything He made was very good (Genesis 1:31). Everything was perfect.

There was no death or suffering, earthquakes, volcanic eruptions, hurricanes, or floods. The climate would have been perfect.

Now this doesn't mean the climate was always the same. Not at all. When God created the sun, moon, and stars on day four of the creation week, He said, *"and let them be for signs and for seasons, and for days and years"* (Genesis 1:14).

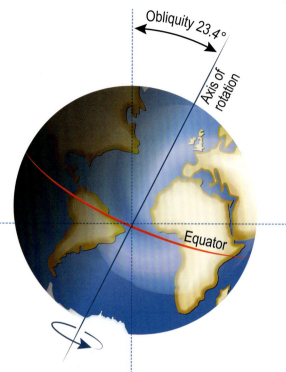

God created the earth tilted on its axis, and no doubt one of the reasons was to create the different seasons. So it certainly seems there must have been seasons in the original world. However, everything about these seasons would have been perfect. God created the earth to be inhabited by man and made sure everything to sustain life on earth was available.

From God's Word

GENESIS 1:31

it was so. ³¹ And God saw everything that he had made, and behold, it was very good. And there was evening and there was morning, the sixth day.

very good!

▶ **There wouldn't have been the extreme seasons we see in parts of the world today with severe cold and intense heat.**

Do you know what one of the most important compounds God made is? This compound is essential for life, and also essential for controlling the earth's climate. This vital compound is water, and it exists in three states: liquid, solid, and gas (water vapor).

On day one of the creation week, when God created the earth, it was covered with water. Now on day two of the creation week, one of the things God did was prepare the atmosphere for life. Then on day three God made the dry land appear and the plants grow. In Genesis 2, we are told God created a system to be part of watering the ground as plants need water: *"and a mist was going up from the land and was watering the whole face of the ground"* (Genesis 2:6).

Water is so important and water vapor is by far the most significant of what we call "greenhouse gases" in our atmosphere. Water dissolves all sorts of substances that are vital for living cells to function and for life to exist. There are 310,000,000 cubic miles of water on the earth's surface that can hold a tremendous amount of heat. And heat can be transferred quickly from water to the air by evaporation and condensation. This helps provide great stability for the earth's temperature and is important for the water cycle that provides needed moisture for living things. Cloud formation (clouds consist of water droplets or ice crystals) is also important for the water cycle and for temperature control. It's all very complex.

Only two people, Adam and Eve, have ever lived on earth with an absolutely perfect climate. We don't know what it's like to live in a perfect world with a perfect climate. We can try to imagine what it must have been like, but we live in a world ruined by sin, so it's impossible for us to totally understand what the perfect world was like when God first made it. Now, because of other events we will learn about, Adam and Eve's descendants until the time of Noah, would have likely experienced a climate not much different from the climate of the perfect world God originally made.

It seems most of our major oil and gas deposits were formed as a result of Noah's Flood burying enormous amounts of animal and plant matter, so people in the world before the Flood probably had very different technology. We can't even imagine what inventions they made, but man has always been highly intelligent, so no doubt there were many marvelous inventions because of man's God-given creative genius. Now much of our technology today exists because of oil and gas, so we do live in a very different world.

▶ **God set all this in place originally to provide the perfect weather for earth and its abundance of life God had created.**

When God created Adam and Eve, He said: *"And let them have dominion over the fish of the sea and over the birds of the heavens and over the livestock and over all the earth and over every creeping thing that creeps on the earth"* (Genesis 1:26).

So God put man in charge of the earth and living things. This means humans could use the earth for man's good and God's glory. In a perfect world before sin, nothing Adam and Eve could have done would have changed the perfect climate patterns God had set up. Oh, if only everything had stayed this way, but sadly it didn't.

From God's Word

GENESIS 1:28

BE GOOD STEWARDS

²⁸ And God blessed them. And God said to them, "Be fruitful and multiply and fill the earth and subdue it, and have dominion over the fish of the sea and over the birds of the heavens and over every living thing that moves on the earth." ²⁹ And God said, "Behold, I have

Groaning
CLIMATE AGE

When God created Adam and Eve, as a test of obedience, He told them they could eat the fruit of all the trees except one. Sadly, they disobeyed and ate that fruit. Their disobedience is called sin, and because all humans are descendants of Adam and Eve, all mankind now has a sin nature.

With sin, everything changed. The earth, universe, and, eventually, the climate were greatly impacted by man's sin. And so was mankind. You see, God judged man's sin with death. Now our bodies age and die. Because we are made in the image of God, our souls, the real us that lives inside our bodies, will live forever. As sinners we couldn't live with a holy God. That's why God promised a Savior (Genesis 3:15) who would come to provide a way for us as sinners to be with God.

Four thousand years after Adam's fall, God's Son stepped into history to be our Savior – the babe in a manger. God's Son, Jesus, died on a Cross for our sins and was raised from the dead so all who receive the free gift of salvation He offers us will go to spend eternity with our Creator in heaven. Heaven will be a perfect place. But right now, because of sin, God's Word tells us that: "For we know that the whole creation has been groaning together in the pains of childbirth until now" (Romans 8:22). The earth, and in fact the whole universe, is not perfect anymore.

From God's Word

ROMANS 8:22

of the children of God. 22 For we know that the whole creation has been groaning together in the pains of childbirth until now. 23 And not

Because Adam was given dominion over the whole of creation, when he rebelled (sinned), the whole creation came under judgment. Therefore everything, including the earth and all life, suffered God's judgment because of sin.

For example, God cursed the ground so it would be more difficult to grow plants and obtain food (Genesis 3:17). In the perfect world, work for man would have always been a joy. But now, God told Adam that work would be tiring and difficult (Genesis 3:19).

Because everything changed as a result of sin, presumably there would be climate change too. No longer would the climate be perfect. God doesn't give us much in the way of details about what may have happened to the climate because of sin. However, we know for sure it would no longer be in perfect harmony as it was in the original perfect creation. But we also can infer that the climate before the Flood certainly wouldn't have had the extremes we see today (those extremes are often created by post-Flood geographical features).

In Colossians, we read of Jesus (who is our Creator), *"And he is before all things, and in him all things hold together"* (Colossians 1:17). The whole universe is held together by the power of God. Before Adam's sin (the fall), God held everything together perfectly. But now, because of sin, God obviously withdrew some of His power that is holding everything together, so now everything runs down. People who reject God's Word and don't believe we are sinners as descendants of Adam and Eve, will not understand that we are living in "a fallen world," no longer held together perfectly by God.

Within this Groaning Climate Age, we are going to learn about some unique events that resulted in sometimes catastrophic climate change. In other words, other events made this groaning much worse.

▶ **The Groaning Climate Age continues to this day, because God has not yet made a new heavens and earth to restore perfection, as He promised in His Word that He will do one day.**

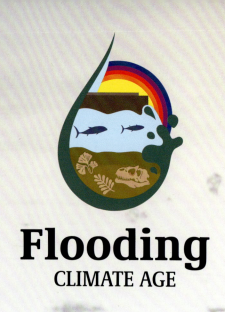

Flooding
CLIMATE AGE

In today's world, there are people who blame flooding on climate change supposedly caused by man. Well, we will cover this when we get to the Shifting Climate Age. But for this Flooding Climate Age, it was the great worldwide Flood that caused massive climate change from which we are really still suffering to this day.

When the first man, Adam, sinned, a spiritual "climate change" began. Humans and all of Creation would no longer be perfect, because of sin. And because of sin, we humans don't want to obey God; we want to be our own gods. As Christians, we need to make sure we don't let sin master us and cause us to do things against what God has instructed.

At the time of Noah, about 4,300 years ago, there was a massive spiritual "climate problem." The whole world, except for Noah and his family, had rebelled against God. As we read in Genesis 6, God said He would destroy the world with a Flood, and commanded Noah to build a great ship (an ark) to accommodate representatives of each of the land-dwelling animal kinds so they would live in the world after the Flood.

This Flood wasn't just a local flood, it was a global (worldwide) cataclysmic flood. The entire earth would be covered in water (see Genesis 7). After the Flood, in Genesis 9, God said that when we see a rainbow, that this would be a sign that He would never again send

From God's Word

GENESIS 7:10

[EVIL PUNISHED]

Noah, as God had commanded Noah. ¹⁰ And after seven days the waters of the flood came upon the earth.

another global flood. Here is what God said about the rainbow:

"I have set my bow in the cloud, and it shall be a sign of the covenant between me and the earth. When I bring clouds over the earth and the bow is seen in the clouds, I will remember my covenant that is between me and you and every living creature of all flesh. And the waters shall never again become a flood to destroy all flesh. When the bow is in the clouds, I will see it and remember the everlasting covenant between God and every living creature of all flesh that is on the earth." God said to Noah, *"This is the sign of the covenant that I have established between me and all flesh that is on the earth."* (Genesis 9:13–17)

While we do see many local floods in our day, never will we see another flood that covers the entire earth as it did in Noah's day. And I suspect there weren't major local floods in the pre-Flood world as we see today.

▶ **The Flood was a major catastrophic event in our earth's history. So much so, we would say we are still seeing the effects of this event in various ways, and the earth is still reeling from all that happened.**

EFFECTS OF
Major Geological Changes

We believe (as Genesis 1:9–10 implies) there was one major continent before the Flood. As part of the processes of the global Flood, this continent was stripped bare and largely destroyed. We believe geologic processes God initiated built new continents that we have today and even those were torn apart during the Flood.

There's lots of evidence of major volcanic activity associated with the Flood and we still experience volcanic eruptions since the Flood. Volcanic eruptions can affect climate.

Major earthquakes rocked the world during the Flood. And now, as the continents are still settling down after the Flood, we experience earthquakes. Earthquakes today can cause massive flooding from tidal waves.

As the Flood drew to an end there were major upheavals as mountains were formed as continental plates collided all over the earth. These mountains were probably much higher than mountains in the pre-Flood world. This would certainly cause climate changes in various areas because of winds, temperature differences, and so on.

The churning floodwaters deposited massive amounts of sediment with billions of plants and animals that became fossils or produced the coal and oil deposits we have today. Much of our technology is built as a result of these deposits.

THE FLOOD
Major Biological Changes

All plant and animal life was wiped off the earth. All land-dwelling animals, except for representatives of each kind saved on Noah's ark, died.

Plants had to repopulate the earth, so it would take time for various plant communities to form and that would depend on the climates in each area.

Because of the effects of the Flood, mountains, changing winds, water temperatures, volcanic debris in the atmosphere, and the Ice Age (as we will learn about next), climates would change over time. Thus, plant communities could change — some lush areas might turn into deserts and some arid areas might become more lush.

Land animals would migrate from where the Ark landed on the mountains of Ararat (somewhere in what we call the Middle East today) to move out over the earth. People also would migrate from the Ark, and then migrate again after the event of the Tower of Babel.

People would clear land to farm, raise animals, and create civilizations. Depending on where people settled, they would experience different climates. And people could affect what happens in an area because of building dams or changing how a river flows, and so on.

▶ **Not only does the world still suffer from the Fall, but the Flood caused major changes for the future.**

69

Icy
CLIMATE AGE

We know that about one-third of the earth's surface was once covered in ice and snow. We see evidence of valleys carved by glaciers and deposits of rocks left by the glaciers as they retreated. Evolutionists believe this is evidence of many ice ages over millions of years. But biblical creationists believe this evidence was left behind by one major ice age generated by the Flood of Noah's day.

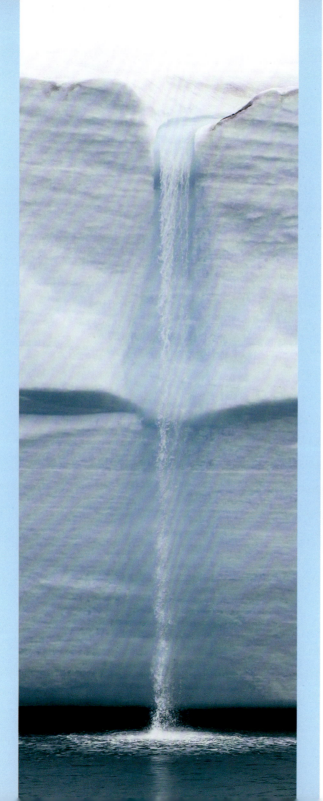

At the end of the Flood, the ocean waters would have been warm from both the land movements and volcanic action pouring hot magma into the oceans. At the same time, the volcanic ash and aerosols in the atmosphere, along with clouds, would have created a cooling effect. Thus there would be a lot of increased evaporation of warm water from the oceans, and because of the cooling effect in the atmosphere, a lot of this water vapor would turn into snow. Vast snowfalls would build up large glaciers of ice.

From God's Word

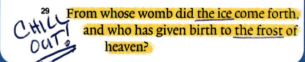

JOB 38:29

29 From whose womb did the ice come forth, and who has given birth to the frost of heaven?

CHILL OUT!

This all had a massive effect on temperatures in any given area. It is believed this ice and snow buildup occurred for several hundred years after the Flood, until things began to settle down.

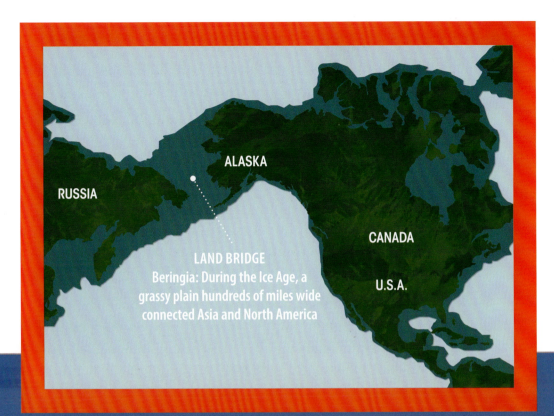

LAND BRIDGE
Beringia: During the Ice Age, a grassy plain hundreds of miles wide connected Asia and North America

▶ **We are still on an earth that's settling down from the catastrophic Flood about 4,300 years ago and the ice age that was generated from it.**

During the time of maximum snow and ice, ocean levels would have been lowered and so there would be land bridges (like across today's Bering Strait) to enable land animals and people to migrate to different parts of the world.

Such a massive ice age would have had dramatic effect on climate worldwide because of ocean currents, winds, and so on. (Of course, that's not all — over the years, scientists have found that sunspot activity and other natural processes also have a big impact on climate. As I said earlier, it's all very complex!).

Now as the earth began to settle down, geologically, the atmosphere would have slowly warmed up as the dust cleared, and the oceans began to cool as the volcanic eruptions and land movements slowed. As a result of this, snowfall lessened, glaciers began to melt back, exposing the various valleys they carved, and huge lakes burst their dams, causing flooding and carving canyons (such as the Grand Canyon). This would also cause climates to shift in various places.

It's important to understand that we will never see all the glaciers that have melted grow back to what they once were because there will never again be conditions created by a global flood.

Shifting
CLIMATE AGE

Now that we understand we live in a fallen, groaning world because of sin, and that the Flood of Noah's day and resulting ice age have caused massive climate changes on earth, we must understand that man can never make changes to make the climate perfect. The fact is, while this earth exists, there will be hurricanes, tornadoes, floods, and all sorts of catastrophic events, and nothing man can do will change this.

We must take note of the promise God made to Noah after the Flood, *"While the earth remains, seedtime and harvest, cold and heat, summer and winter, day and night, shall not cease"* (Genesis 8:22). God set up this earth to cope with all the changes that occur because of sin and the changes caused by the Flood and ice age. No matter what man does, man can never destroy the earth! Only God can (and will) do that.

Now don't get me wrong! This doesn't mean man can do anything he wants to this earth. Remember, God gave man dominion over the creation. But that doesn't mean authority to abuse it. However, it does mean that man is to use it for man's good and God's glory.

From God's Word

JOB 38:26–27

[God is in control]

26 to bring rain on a land where no man is, on the desert in which there is no man, 27 to satisfy the waste and desolate land, and to make the ground sprout with grass?

Sinful man thinks everything natural is good. But everything suffers from the fall of man and the ground is cursed: *"cursed is the ground because of you; in pain you shall eat of it all the days of your life; thorns and thistles it shall bring forth for you"* (Genesis 3:17–18).

Only people are made in the image of God. Humans are not just animals. Human beings have immortal souls that will live forever. God sent His Son to pay the penalty for sin, so if we receive His free gift of salvation, we will live forever with our Creator. This means we shouldn't value animals above humans. And if we can use things on this earth, or change things to benefit humans, then we should do this in the best way that glorifies God and also looks after the creation God entrusted to us.

So the bottom line is, climate change on this earth will not stop until Jesus returns. Now I've had people say to me, "But look at all those devastating tornadoes in America, isn't that because of climate change?" I tell them to do the research and you'll find there were worse tornadoes and more of them in the past. So yes, all this is because of climate change, but climate change because of this fallen world we live in. Others have said "We've never seen so many people affected by flooding in Australia or America!" But then I remind them that there's a lot more people living now and living in places that have not been developed for housing before.

▶ **Much of the modern environmental movement involves giving the creation dominion over man. This climate change activism is a religion that worships nature and man instead of God.**

It seems today that whenever there's news of a hurricane, cyclone, tornado, or flood we will hear many politicians claim it's because of man-made climate change. Making such statements without have the real data and being truly scientific about it should make us realize even more that much of this climate alarmism is a religion centered on man and our perceived ability to save ourselves – really worship of man and the environment. It reminds me of Romans 1:23: *"and exchanged the glory of the immortal God for images resembling mortal man and birds and animals and creeping things."*

Fiery
CLIMATE AGE

Think back to the promise God made in Genesis 8:22: *"While the earth remains, seedtime and harvest, cold and heat, summer and winter, day and night, shall not cease."* When you hear someone today claim that man will destroy the earth in a few years, you can know one hundred percent for sure this is not true. Man cannot destroy the earth. God promised that. But God's Word makes it clear that one day, Jesus, who returned to the Father in Heaven after His Resurrection, will one day return to this fallen, groaning earth and judge this earth and the whole universe with a fiery end! But He will make a new heavens and earth as we will discuss in the final climate age.

It's interesting that in 2 Peter 3:3–7, we read that people in the last days (and we've been living in the last days since Jesus came to earth as the babe in a manger to be our Savior) will reject that God created the earth, reject the Flood of Noah's day, and reject the coming judgment by fire:

"Knowing this first of all, that scoffers will come in the last days with scoffing, following their own sinful desires. They will say, 'Where is the promise of his coming? For ever since the fathers fell asleep, all things are continuing as they were from the beginning of creation.' For they deliberately overlook this fact, that the heavens existed long ago, and the earth was formed out of water and through water by the word of God, and that by means of these the world that then existed was deluged with water and perished. But by the same word the heavens and earth that now exist are stored up for fire, being kept until the day of judgment and destruction of the ungodly."

From God's Word

2 PETER 3:10

FIRE → NEW EARTH!

should reach repentance. ¹⁰ But the day of the Lord will come like a thief, and then the heavens will pass away with a roar, and the heavenly bodies⁷ will be burned up and dissolved, and the earth and the works that are done on it will be exposed.⁸

Now think about this. Those who have adopted the climate religion of our day reject that God created the earth. They believe everything evolved by chance and random processes over millions of years. They certainly reject that man is a sinner and the earth is groaning because of sin. They reject the Flood of Noah's day (and the resulting ice age), and so reject one of the significant causes of major climate change today. And they reject the coming fiery judgment of the earth by God.

They believe man must save the planet before we destroy it. They believe man can save himself. But in 2 Peter 3:17–18, God also tells us that, *"You therefore, beloved, knowing this beforehand, take care that you are not carried away with the error of lawless people and lose your own stability. But grow in the grace and knowledge of our Lord and Savior Jesus Christ. To him be the glory both now and to the day of eternity."*

▶ Make sure you are listening to God's Word to truly understand life and the universe.

Don't be led astray by these people who reject the truth of God's Word. Mankind cannot save itself. Most of all, make sure you understand the problem we humans have (sin) and what God has said He did to save us from our sin. And make sure you have received that free gift of salvation He offers to us. This leads us to the final climate age.

Heavenly
CLIMATE AGE

Around 6,000 years ago, God created the universe and all life on earth. He created two humans who rebelled against Him and brought sin and death into the world. As all humans are descendants of these two people, Adam and Eve, all are sinners, *"for all have sinned and fall short of the glory of God"* (Romans 3:23). And because we are all sinners, all are under the judgment of death, *"Therefore, just as sin came into the world through one man, and death through sin, and so death spread to all men because all sinned"* (Romans 5:12).

Because humans are made in God's image, we have a soul that will live forever even though our bodies die. But a sinner can't live with a holy God, which is why God has provided a way for us to live with Him for eternity, *"For God so loved the world, that he gave his only Son, that whoever believes in him should not perish but have eternal life"* (John 3:16).

"If you confess with your mouth that Jesus is Lord and believe in your heart that God raised him from the dead, you will be saved" (Romans 10:9). And salvation is a gift from God; there is nothing we can do to earn this: *"For by grace you have been saved through faith. And this is not your own doing; it is the gift of God"* (Ephesians 2:8). Now for those who have received this free gift, when our bodies die, we as a person do not die, but go to be with the Lord forever. However those who reject this free gift will suffer eternal separation from God that He calls a "second death": *"But as for the cowardly, the faithless, the detestable, as for murderers, the sexually immoral, sorcerers, idolaters, and all liars, their portion will be in the lake that burns with fire and sulfur, which is the second death"* (Revelation 21:8).

Now let's return to the passage in 2 Peter 3 where we learn that these scoffers who reject that God created

▶ **God wants these scoffers to repent of their sin.
He wants them to receive the free gift of salvation.**

and reject the Flood, scoff that Jesus is going to return because they believe things will just go on and on into the future.

They reject the coming judgment by fire. Now listen to what God says in His Word as to why Jesus hasn't returned yet:

"But do not overlook this one fact, beloved, that with the Lord one day is as a thousand years, and a thousand years as one day. The Lord is not slow to fulfill his promise as some count slowness, but is patient toward you, not wishing that any should perish, but that all should reach repentance" (2 Peter 3:8–9).

So the next time you see people trapped in this climate change religion claiming that man is going to destroy the earth because of climate change, pray for them. Pray that they will stop worshiping man and worshiping the earth. And pray that they would have a spiritual "climate change" in their hearts and repent of their sin and receive the free gift of salvation God offers through His Son the Lord Jesus Christ who died on that Cross 2,000 years ago and was raised from the dead paying sin's penalty.

And because we know it is God who will destroy the earth to create a new one in His timing at some point in the future, let's be joyful as we wait for that wonderful day.

From God's Word

2 PETER 3:13

GOD'S PROMISES KEPT!

will melt as they burn! ¹³ But according to his promise we are waiting for new heavens and a new earth in which righteousness dwells.

"Since all these things are thus to be dissolved, what sort of people ought you to be in lives of holiness and godliness, waiting for and hastening the coming of the day of God, because of which the heavens will be set on fire and dissolved, and the heavenly bodies will melt as they burn! But according to his promise we are waiting for new heavens and a new earth in which righteousness dwells" (2 Peter 3:11–13).

Stand Firm

A strong biblical worldview is vital to living in a climate-changed world. Neither people nor science have all the answers to be able to effectively understand the changes we experience on earth. God is essential to our understanding of climate change and our future. Don't be confused or frightened – you know the history and importance of the seven climate ages. Remember His promise in Genesis 8:22.

Perfect CLIMATE AGE | **Groaning** CLIMATE AGE | **Flooding** CLIMATE AGE | **Icy** CLIMATE AGE | **Shifting** CLIMATE AGE | **Fiery** CLIMATE AGE | **Heavenly** CLIMATE AGE

*"While the earth remains,
seedtime and harvest,
cold and heat,
summer and winter,
day and night, shall not cease."*
Genesis 8:22

Answers in Genesis has other resources detailing how to build a biblical worldview of environmentalism, including many articles on the AnswersinGenesis.org website.

Endnotes

1. Richard Wilson, "Power Policy – Plan or Panic?" Science and Public Affairs, Bulletin of the Atomic Scientists, May 1972, Volume XXVIII, Number 5; pages 29 and 30. https://books.google.com/books?id=pwsAAAAAMBAJ&lpg=PA29&ots=zoY4QTY67Z&dq=%22u.s.%20oil%20supplies%20will%20last%22&pg=PA29#v=onepage&q=%22u.s.%20oil%20supplies%20will%20last%22&f=false; "Ice Sheets, Rising Seas, Floods," The Discovery of Global Warming, May 2023. https://history.aip.org/climate/floods.htm; "Feb. 12, 1958: Popular TV Program Warns of Global Warming," Zinn Educational Project, This Day in History. https://www.zinnedproject.org/news/tdih/popular-tv-program-warns-of-global warming/#:~:text=On%20February%2012%2C%201958%2C%20the,catastrophic%20rise%20in%20sea%20levels; Peter Gwynne, "The Cooling World," Newsweek, April 28, 1975; available online at http://denisdutton.com/newsweek_coolingworld.pdf; Chapter 16, "Climate Change Facts: Should We Be Concerned." The New Answers Book 4 by Dr. Alan White on October 2, 2013; https://answersingenesis.org/environmental-science/climate-change/should-we-be-concerned-about-climate-change/; "Warming or Cooling?" Timeline Milestones, "The Discovery of Global Warming, May 2023, https://history.aip.org/climate/timeline.htm; 1985 Villach Conference; Conference Statement. https://library.wmo.int/viewer/28228?medianame=wmo_661_en_#page=11&viewer=picture&o=bookmarks&n=0&q=; "The origins of the IPCC: How the world woke up to Climate Change." International Science Council Blog, October 3, 2018. https://council.science/current/blog/the-origins-of-the-ipcc-how-the-world-woke-up-to-climate-change/; Ken Allen, "Global Warming and the Sun," Jan. 2000, RECORDER Vol. 25, NO. 01. https://csegrecorder.com/articles/view/global-warming-and-the-sun; Larry Vardiman, "New Evidence for Global Cooling," Institute for Creation Research, http://www.icr.org/article/new-evidence-for-global-cooling/; and Larry Vardiman, "Will Solar Inactivity Lead to Global Cooling?" Institute for Creation Research, http://www.icr.org/article/will-solar-inactivity-lead-global-cooling/; Theodor Landscheidt, "Solar Activity: A Dominant Factor In Climatic Dynamics;" Energy & Environment. Vol. 9, No. 6 (September 1998). Energy & Environment. https://www.jstor.org/stable/44396937; Bethan Davies. "Coral Reefs in Protected Areas can Recover Quickly After Mass Coral Bleaching Events." March 25, 2022. https://www.azocleantech.com/news.aspx?newsID=31434; Holly P. Jones, and Oswald J. Schmitz, "Rapid Recovery of Damaged Ecosystems," PLoS One. May 27, 2009. https://www.ncbi.nlm.nih.gov/pmc/articles/PMC2680978/; Dr. Alan White, "Chapter 16 – Climate Change Facts: Should We Be Concerned?" Answers October. https://answersingenesis.org/environmental-science/climate-change/should-we-be-concerned-about-climate-change/; "Climate Models and the Hiatus in Global Mean Surface Warming of the Past 15 Years" and "TS.2.2.1 Changes in Temperature; Surface." Climate Change 2013, The Physical Science Basis, IPCC. p. 37, 61.

2. "Daylight, Darkness and Changing of the Seasons at the North Pole", NOAA. https://www.pmel.noaa.gov/arctic-zone/gallery_np_seasons.html#:~:text=The%20North%20Pole%20stays%20in,to%20sink%20towards%20the%20horizon

3. IPCC. (2022), Framing and Context. In Global Warming of 1.5°C: IPCC Special Report on Impacts of Global Warming of 1.5°C above Pre-industrial Levels in Context of Strengthening Response to Climate Change, Sustainable Development, and Efforts to Eradicate Poverty (p. 49–92). Cambridge: Cambridge University Press.

4. "Missions: Earth Observing Systems (EOS)" NASA. https://eospso.nasa.gov/mission-category/3; https://eospso.nasa.gov/missions/nimbus-1; https://eospso.nasa.gov/missions/active-cavity-radiometer-irradiance-monitor-satellite

5. Roy Spencer and John Christy. (2022). Dependence of Climate Sensitivity Estimates on Internal Climate Variability During 1880–2020. 10.21203/rs.3.rs-2162757/v1.

6. R. McKitrick and J. Christy, (2020). Pervasive warming bias in CMIP6 tropospheric layers. Earth and Space Science, 7, e2020EA001281. https://doi.org/10.1029/2020EA001281.

7. EPA.Overview of Greenhouse Gases. https://www.epa.gov/ghgemissions/overview-greenhouse-gases.

8. Rod Martin. "A Proposed Bible-Science Perspective on Global Warming." Answers Research Journal 3 (2010): 91–106. https://answersresearchjournal.org/bible-science-global-warming/.

9. ACS 2022. American Chemical Society National Historic Chemical Landmarks. The Keeling Curve. http://www.acs.org/content/acs/en/education/whatischemistry/landmarks/keeling-curve.html (August, 2022).

10. Craig D. Idso, 2013. The Positive Externalities of Carbon Dioxide: Estimating the Monetary Benefits of Rising Atmospheric CO2 Concentrations on Global Food Production. http://www.co2science.org/education/reports/co2benefits/.

11. Ibid.

12. NOAA, "What are El Niño and La Niña?" https://oceanservice.noaa.gov/facts/ninonina.html

13. Ibid.

14. NOAA, U.S. Hurricane Strikes by Decade. https://www.nhc.noaa.gov/pastdec.shtml

15. John R. Christy. Examination of Extreme Rainfall Events in Two Regions of the United States since the 19th Century. AIMS Environmental Science, 2019, 6(2): 109–126. doi: 10.3934/environsci.2019.2.109.

16. Gordon Wilson, "Wildfires: Cause for Alarm?" In Answers Magazine, April 1, 2021.

17. Ibid.

18. Myron Ebell, "Australian Wildfires Were Caused by Humans, Not Climate Change," Competitive Enterprise Institute, January 8, 2020

19. A Proposed Bible-Science Perspective on Global Warming, by Rod Martin, published on May 26, 2010. Answers Research Journal 3 (2010): 91–106. https://answersresearchjournal.org/bible-science-global-warming/.

20. Michael J. Oard, "Setting the Stage for an Ice Age" in Answers Magazine, https://answersingenesis.org/environmental-science/ice-age/setting-the-stage-for-an-ice-age/.

21. Michael Oard, "How Much Global Warming Is Natural?" Answers in Depth, https://answersingenesis.org/environmental-science/climate-change/how-much-global-warming-is-natural/.

22. Michael J. Oard, "Setting the Stage for an Ice Age" in Answers Magazine, https://answersingenesis.org/environmental-science/ice-age/setting-the-stage-for-an-ice-age/.

23. Andrew B. Appleby, "Epidemics and Famine in the Little Ice Age," The Journal of interdisciplinary history 10, no. 4 (1980): 643–663.

24. Michael Oard, "How Much Global Warming Is Natural? Answers in Depth, https://answersingenesis.org/environmental-science/climate-change/how-much-global-warming-is-natural/.

25. Vijay Jayaraj, "No Reason to Panic: The State of Arctic Ice Mass And Greenland," Cornwall Alliance, https://cornwallalliance.org/2019/09/no-reason-to-panic-the-state-of-arctic-ice-mass-and-greenland/ Institute of Polar Research https://ads.nipr.ac.jp/vishop/#/extent; National Snow and Ice Data Center https://nsidc.org/arcticseaicenews/charctic-interactive-sea-ice-graph/.

26. Vijay Jayaraj, "Arctic Ice at Decade High Level," Cornwall Alliance, https://cornwallalliance.org/2022/09/arctic-ice-at-decade-high-level-can-doomsayers-explain.

27. https://www.arcticwwf.org/wildlife/polar-bear/; Jon Miltimore, "The Myth that the Polar Bear Population is Declining," Cornwall Alliance. https://cornwallalliance.org/2019/09/the-myth-that-the-polar-bear-population-is-declining/.

28. National Oceanic and Atmospheric Administration, "Relative Sea Level Trend, 8518750 The Battery, New York," Tides and Currents, tidesandcurrents.noaa.gov, accessed August 13, 2021.

29. NOAA. "What is Coral Bleaching?" https://oceanservice.noaa.gov/facts/coral_bleach.html

30. "What Are Coral Reefs." Coral Reef Information System. https://www.coris.noaa.gov/about/what_are/#:~:text=Comprising%20over%206%2C000%20known%20species,and%20building%20up%2C%20reef%20structures.

31. "Highest coral cover in central, northern reef in 36 years"; Australian Institute of Marine Science; https://www.aims.gov.au/information-centre/news-and-stories/highest-coral-cover-central-northern-reef-36-years; Australia's Tropical Marine Research Agency. AIMS. (2022, November 29). Retrieved January 8, 2023, from https://www.aims.gov.au/

32. The New York Times, "Amid Heat Wave California Asks Electric Vehicle Owners to Limit Charging," https://www.nytimes.com/2022/09/01/us/california-heat-wave-flex-alert-ac-ev-charging.html.

33. "Will "Carbon Cuts" Make Climate Change Worse?" Ken Ham, Ken Ham Blog, June 12, 2023; https://answersingenesis.org/environmental-science/climate-change/will-carbon-cuts-make-climate-change-worse/

34. "Did Earth's Temperature Break Records This Summer?" Ken Ham, Ken Ham Blog. July 13, 2023; https://answersingenesis.org/environmental-science/did-earths-temperature-break-records-this-summer/

35. "Is This Burning Drowning World No Place for More Kids?" Ken Ham, Ken Ham Blog. December 7, 2023; https://answersingenesis.org/worldview/is-world-no-place-for-more-kids/

36. Definition of "Climate change." https://www.merriam-webster.com/dictionary/climate%20change#:~:text=noun,Earth's%20climate%20and%20weather%20patterns

Image Credits

L = left, T= top, TL = top left, B=bottom, BL = bottom left, C = center, CR = center right, CL = center left, R = right, TR = top right, BR = bottom right, BC = bottom center

All images are public domain (PD-US, and PD-Art), except for:

Shutterstock.com: p 2-3, p 4 BL, BC, p 5 BC, BR, p 6, B, p 6-7, p 8-9, p 10 TR, p 11 all, p 12-13, p 13 all, p 14, C x2, B x3, p 15 all, CL,p 18 T, p 18-19, p 20-21 all, p 22-23 all, p 24-25 all, p 27, bkg, p 31 BL, p 32 B, p 40 BL, p 41 BR, p 42-43, p 42 BL, p 44 BL, p 45 TR, p 45 BR, p 46 CL, TR, p 47 T, B, p 48 BL, p 48-49, p 49 BR, p 50, p 51 all, p 52, p 53, p 56-57 all, p 58-59 bkg, p 59 R, p 60 all, p 61, p 62-63 B, p 63 all, p 64, p 65, p 66, p 67, p 68-69 C, p 70, p 72, p 73, p 74 BR, p 75, p 76 B, p 77 R, p 78-79 bkg.

Shutterstock AI: cover

istock.com: p 35 R, p 41 T

flickr: p 29 R (USDA photo by Bob Nichols), p 68 L, Anthony Quintano,

Superstock.com: p 4-5,

USGS: p 32 T,

UNESCO: p 26-27 C, p 34 all, p 35 L, B,

NASA: 14 CR, p 17 BR all satellites, p 43 BL,

NOAA: p 17 BL, p 27 BC, BL, p 33 C & B, p 36 T, p 42 BC

US Forestry Service: p 30 Brandon Dethlefs, p 31 TL

Ark Encounter: 43 BR, p 54-55, p 69 T, p 71 T,

NLPG staff: Climate Age logos

Wikimedia Commons: p 28 T, p 29 L, p 37 Plane, Brandon Dethlefs, P 39 R, p 40 T, Ansgar Walk, p 42 BR p 44 T, p 68 BL, Hansueli Krapf
Images from Wikimedia Commons are used under the CC0 1.0, CC BY-SA 2.0 DE, CC-BY-SA-3.0 license or the GNU Free Documentation License, Version 1.3.
Wiki/4.0 international/TomaszSwatowski: p 16 B.
Wiki/4.0/Geostationary Operational Environmental Satellite Program: pg 27 L
Wiki/4.0 International/Helioseismic and Magnetic Imagery (HMI): p 38